Omnilingual

by

H. Beam Piper

Double 9 BOOKS

OMNILINGUAL
by H. BEAM PIPER

ISBN: 978-93-59324-42-5

Published by

DOUBLE 9 BOOKS
2/13-B, Ansari Road, Daryaganj
New Delhi – 110002
info@double9books.com
www.double9books.com
Tel. 011-40042856

ABOUT THE AUTHOR

American science fiction author Henry Beam Piper lived from March 23, 1904, through November 6, 1964. His lengthy Terro-Human Future History trilogy and a more condensed collection of "Paratime" alternate history novellas are his most well-known works. His name is listed as "Horace Beam Piper" in another source, along with a different death date. "Henry Beam Piper," it states on his tombstone. Piper may have contributed to some of the confusion when he stated that the H stood for Horace, leading some to believe that he did it because he disliked his name. The majority of Piper's education came through self-learning; he did not "submit myself to the absurd pain of four years in the unpleasant constraints of a raccoon coat" in order to learn science and history. He started working as a worker at the Pennsylvania Railroad's Altoona yards in Pennsylvania when he was 18 years old. He also worked for the railroad as a night watchman. When Piper's career appeared to be in trouble in 1964, he killed himself because he was afraid to ask for help and because he adhered to libertarian principles. The last entry in his diary was dated November 5, and his Pennsylvania death certificate states that his body was discovered on November 8. The precise date of his passing is unknown.

Martha Dane paused, looking up at the purple-tinged copper sky. The wind had shifted since noon, while she had been inside, and the dust storm that was sweeping the high deserts to the east was now blowing out over Syrtis. The sun, magnified by the haze, was a gorgeous magenta ball, as large as the sun of Terra, at which she could look directly. Tonight, some of that dust would come sifting down from the upper atmosphere to add another film to what had been burying the city for the last fifty thousand years.

The red loess lay over everything, covering the streets and the open spaces of park and plaza, hiding the small houses that had been crushed and pressed flat under it and the rubble that had come down from the tall buildings when roofs had caved in and walls had toppled outward. Here, where she stood, the ancient streets were a hundred to a hundred and fifty feet below the surface; the breach they had made in the wall of the building behind her had opened into the sixth story. She could look down on the cluster of prefabricated huts and sheds, on the brush-grown flat that had been the waterfront when this place had been a seaport on the ocean that was now Syrtis Depression; already, the bright metal was thinly coated with red dust. She thought, again, of what clearing this city would mean, in terms of time and labor, of people and supplies and equipment brought across fifty million miles of space. They'd have to use machinery; there was no other way it could be done. Bulldozers and power shovels and draglines; they were fast, but they were rough and indiscriminate. She remembered the digs around Harappa and Mohenjo-Daro, in the Indus Valley, and the careful, patient native laborers—the painstaking foremen, the pickmen and spademen, the long files of basketmen carrying away the earth. Slow and primitive as the civilization whose ruins they were uncovering, yes, but she could count on the fingers of one hand the times one of her pickmen had damaged a valuable object in the ground. If it hadn't been for the underpaid and uncomplaining native laborer, archaeology would still be back where Wincklemann had found it. But on Mars there was no native labor; the last Martian had died five hundred centuries ago.

Something started banging like a machine gun, four or five hundred yards to her left. A solenoid jack-hammer; Tony Lattimer must have decided which building he wanted to break into next. She became conscious, then, of the awkward weight of her equipment, and began redistributing it, shifting the straps of her oxy-tank pack, slinging the camera from one shoulder and the board and drafting tools from the other, gathering the notebooks and sketchbooks under her left arm. She started walking down the road, over hillocks of buried rubble, around snags of wall jutting up out of the loess, past buildings still standing, some of them already breached and explored, and across the brush-grown flat to the huts.

There were ten people in the main office room of Hut One when she entered. As soon as she had disposed of her oxygen equipment, she lit a cigarette, her first since noon, then looked from one to another of them. Old Selim von Ohlmhorst, the Turco-German, one of her two fellow archaeologists, sitting at the end of the long table against the farther wall, smoking his big curved pipe and going through a looseleaf notebook. The girl ordnance officer, Sachiko Koremitsu, between two droplights at the other end of the table, her head bent over her work. Colonel Hubert Penrose, the Space Force CO, and Captain Field, the intelligence officer, listening to the report of one of the airdyne pilots, returned from his afternoon survey flight. A couple of girl lieutenants from Signals, going over the script of the evening telecast, to be transmitted to the *Cyrano*, on orbit five thousand miles off planet and relayed from thence to Terra via Lunar. Sid Chamberlain, the Trans-Space News Service man, was with them. Like Selim and herself, he was a civilian; he was advertising the fact with a white shirt and a sleeveless blue sweater. And Major Lindemann, the engineer officer, and one of his assistants, arguing over some plans on a drafting board. She hoped, drawing a pint of hot water to wash her hands and sponge off her face, that they were doing something about the pipeline.

She started to carry the notebooks and sketchbooks over to where Selim von Ohlmhorst was sitting, and then, as she always did, she turned aside and stopped to watch Sachiko. The Japanese girl was restoring what had been a book, fifty thousand years ago; her eyes were masked by a binocular loup, the black headband invisible against her glossy black hair, and she was picking delicately at the crumbled page with a hair-fine wire set in a handle of copper tubing. Finally, loosening a particle as tiny as a snowflake, she grasped it with tweezers, placed it on the sheet of transparent plastic on which she was reconstructing the page, and set it with a mist of fixative from a little spraygun. It was a sheer joy to watch her; every movement was as graceful and precise as though done to music after being rehearsed a hundred times.

"Hello, Martha. It isn't cocktail-time yet, is it?" The girl at the table spoke without raising her head, almost without moving her lips, as though she were afraid that the slightest breath would disturb the flaky stuff in front of her.

"No, it's only fifteen-thirty. I finished my work, over there. I didn't find any more books, if that's good news for you."

Sachiko took off the loup and leaned back in her chair, her palms cupped over her eyes.

"No, I like doing this. I call it micro-jigsaw puzzles. This book, here, really is a mess. Selim found it lying open, with some heavy stuff on top of it; the pages were simply crushed. " She hesitated briefly. "If only it would mean something, after I did it."

There could be a faintly critical overtone to that. As she replied, Martha realized that she was being defensive.

"It will, some day. Look how long it took to read Egyptian hieroglyphics, even after they had the Rosetta Stone."

Sachiko smiled. "Yes. I know. But they did have the Rosetta Stone."

"And we don't. There is no Rosetta Stone, not anywhere on Mars. A whole race, a whole species, died while the first Crò-Magnon cave-artist was daubing pictures of reindeer and bison, and across fifty thousand years and fifty million miles there was no bridge of understanding.

"We'll find one. There must be something, somewhere, that will give us the meaning of a few words, and we'll use them to pry meaning out of more words, and so on. We may not live to learn this language, but we'll make a start, and some day somebody will."

Sachiko took her hands from her eyes, being careful not to look toward the unshaded light, and smiled again. This time Martha was sure that it was not the Japanese smile of politeness, but the universally human smile of friendship.

"I hope so, Martha: really I do. It would be wonderful for you to be the first to do it, and it would be wonderful for all of us to be able to read what these people wrote. It would really bring this dead city to life again." The smile faded slowly. "But it seems so hopeless."

"You haven't found any more pictures?"

Sachiko shook her head. Not that it would have meant much if she had. They had found hundreds of pictures with captions; they had never been able to establish a positive relationship between any pictured object and any printed word. Neither of them said anything more, and after a moment Sachiko replaced the loup and bent her head forward over the book.

Selim von Ohlmhorst looked up from his notebook, taking his pipe out of his mouth.

"Everything finished, over there?" he asked, releasing a puff of smoke.

"Such as it was." She laid the notebooks and sketches on the table. "Captain Gicquel's started airsealing the building from the fifth floor down, with an entrance on the sixth; he'll start putting in oxygen generators as soon as that's done. I have everything cleared up where he'll be working."

Colonel Penrose looked up quickly, as though making a mental note to attend to something later. Then he returned his attention to the pilot, who was pointing something out on a map.

Von Ohlmhorst nodded. "There wasn't much to it, at that," he agreed. "Do you know which building Tony has decided to enter next?"

"The tall one with the conical thing like a candle extinguisher on top, I think. I heard him drilling for the blasting shots over that way."

"Well, I hope it turns out to be one that was occupied up to the end."

The last one hadn't. It had been stripped of its contents and fittings, a piece of this and a bit of that, haphazardly, apparently over a long period of time, until it had been almost gutted. For centuries, as it had died, this city had been consuming itself by a process of auto-cannibalism. She said something to that effect.

"Yes. We always find that—except, of course, at places like Pompeii. Have you seen any of the other Roman cities in Italy?" he asked. "Minturnae, for instance? First the inhabitants tore down this to repair that, and then, after they had vacated the city, other people came along and tore down what was left, and burned the stones for lime, or crushed them to mend roads, till there was nothing left but the foundation traces. That's where we are fortunate; this is one of the places where the Martian race perished, and there were no barbarians to come later and destroy what they had left." He puffed slowly at his pipe. "Some of these days, Martha, we are going to break into one of these buildings and find that it was one in which the last of these people died. Then we will learn the story of the end of this civilization."

And if we learn to read their language, we'll learn the whole story, not just the obituary. She hesitated, not putting the thought into words. "We'll find that, sometime, Selim," she said, then looked at her watch. "I'm going to get some more work done on my lists, before dinner."

For an instant, the old man's face stiffened in disapproval; he started to say something, thought better of it, and put his pipe back into his mouth. The brief wrinkling around his mouth and the twitch of his white mustache had been enough, however; she knew what he was thinking. She was wasting time and effort, he believed; time and effort belonging not to herself but to the expedition. He could be right, too, she realized. But he had to be wrong;

there had to be a way to do it. She turned from him silently and went to her own packing-case seat, at the middle of the table.

Photographs, and photostats of restored pages of books, and transcripts of inscriptions, were piled in front of her, and the notebooks in which she was compiling her lists. She sat down, lighting a fresh cigarette, and reached over to a stack of unexamined material, taking off the top sheet. It was a photostat of what looked like the title page and contents of some sort of a periodical. She remembered it; she had found it herself, two days before, in a closet in the basement of the building she had just finished examining.

She sat for a moment, looking at it. It was readable, in the sense that she had set up a purely arbitrary but consistently pronounceable system of phonetic values for the letters. The long vertical symbols were vowels. There were only ten of them; not too many, allowing separate characters for long and short sounds. There were twenty of the short horizontal letters, which meant that sounds like -ng or -ch or -sh were single letters. The odds were millions to one against her system being anything like the original sound of the language, but she had listed several thousand Martian words, and she could pronounce all of them.

And that was as far as it went. She could pronounce between three and four thousand Martian words, and she couldn't assign a meaning to one of them. Selim von Ohlmhorst believed that she never would. So did Tony Lattimer, and he was a great deal less reticent about saying so. So, she was sure, did Sachiko Koremitsu. There were times, now and then, when she began to be afraid that they were right.

The letters on the page in front of her began squirming and dancing, slender vowels with fat little consonants. They did that, now, every night in her dreams. And there were other dreams, in which she read them as easily as English; waking, she would try desperately and vainly to remember. She blinked, and looked away from the photostatted page; when she looked back, the letters were behaving themselves again. There were three words at the top of the page, over-and-underlined, which seemed to be the Martian method of capitalization. *Masthamorvod Tadavas Sornhulva.* She pronounced them mentally, leafing through her notebooks to see if she had encountered them before, and in what contexts. All three were listed. In addition, *masthar* was a fairly common word, and so was *norvod,* and so was *nor,* but *-vod* was a suffix and nothing but a suffix. *Davas,* was a word, too, and *ta-* was a common prefix; *sorn* and *hulva* were both common words. This language, she had long ago decided, must be something like German; when the Martians had needed a new word, they had just pasted a couple of existing words together. It would probably turn out to be a grammatical horror. Well, they had published magazines, and one of them had been called *Mastharnorvod Tadavas Sornhulva*. She wondered if it had been something like the *Quarterly Archaeological Review*, or something more on the order of *Sexy Stories*.

A smaller line, under the title, was plainly the issue number and date; enough things had been found numbered in series to enable her to identify the numerals and determine that a decimal system of numeration had been used. This was the one thousand and seven hundred and fifty-fourth issue, for Doma, 14837; then Doma must be the name of one of the Martian months. The word had turned up several times before. She found herself puffing furiously on her cigarette as she leafed through notebooks and piles of already examined material.

Sachiko was speaking to somebody, and a chair scraped at the end of the table. She raised her head, to see a big man with red hair and a red face, in Space Force green, with the single star of a major on his shoulder, sitting down. Ivan Fitzgerald, the medic. He was lifting weights from a book similar to the one the girl ordnance officer was restoring.

"Haven't had time, lately," he was saying, in reply to Sachiko's question. "The Finchley girl's still down with whatever it is she has, and it's something I haven't been able to diagnose yet. And I've been checking on bacteria cultures, and in what spare time I have, I've been dissecting specimens for Bill Chandler. Bill's finally found a mammal. Looks like a lizard, and it's only four inches long, but it's a real warm-blooded, gamogenetic, placental, viviparous mammal. Burrows, and seems to live on what pass for insects here."

"Is there enough oxygen for anything like that?" Sachiko was asking.

"Seems to be, close to the ground." Fitzgerald got the headband of his loup adjusted, and pulled it down over his eyes. "He found this thing in a ravine down on the sea bottom—Ha, this page seems to be intact; now, if I can get it out all in one piece—"

He went on talking inaudibly to himself, lifting the page a little at a time and sliding one of the transparent plastic sheets under it, working with minute delicacy. Not the delicacy of the Japanese girl's small hands, moving like the paws of a cat washing her face, but like a steam-hammer cracking a peanut. Field archaeology requires a certain delicacy of touch, too, but Martha watched the pair of them with envious admiration. Then she turned back to her own work, finishing the table of contents.

The next page was the beginning of the first article listed; many of the words were unfamiliar. She had the impression that this must be some kind of scientific or technical journal; that could be because such publications made up the bulk of her own periodical reading. She doubted if it were fiction; the paragraphs had a solid, factual look.

At length, Ivan Fitzgerald gave a short, explosive grunt.

"Ha! Got it!"

She looked up. He had detached the page and was cementing another plastic sheet onto it.

"Any pictures?" she asked.

"None on this side. Wait a moment." He turned the sheet. "None on this side, either." He sprayed another sheet of plastic to sandwich the page, then picked up his pipe and relighted it.

"I get fun out of this, and it's good practice for my hands, so don't think I'm complaining," he said, "but, Martha, do you honestly think anybody's ever going to get anything out of this?"

Sachiko held up a scrap of the silicone plastic the Martians had used for paper with her tweezers. It was almost an inch square.

"Look; three whole words on this piece," she crowed. "Ivan, you took the easy book."

Fitzgerald wasn't being sidetracked. "This stuff's absolutely meaningless," he continued. "It had a meaning fifty thousand years ago, when it was written, but it has none at all now."

She shook her head. "Meaning isn't something that evaporates with time," she argued. "It has just as much meaning now as it ever had. We just haven't learned how to decipher it."

"That seems like a pretty pointless distinction," Selim von Ohlmhorst joined the conversation. "There no longer exists a means of deciphering it."

"We'll find one." She was speaking, she realized, more in self-encouragement than in controversy.

"How? From pictures and captions? We've found captioned pictures, and what have they given us? A caption is intended to explain the picture, not the picture to explain the caption. Suppose some alien to our culture found a picture of a man with a white beard and mustache sawing a billet from a log. He would think the caption meant, 'Man Sawing Wood.' How would he know that it was really 'Wilhelm II in Exile at Doorn?'"

Sachiko had taken off her loup and was lighting a cigarette.

"I can think of pictures intended to explain their captions," she said. "These picture language-books, the sort we use in the Service—little line drawings, with a word or phrase under them."

"Well, of course, if we found something like that," von Ohlmhorst began.

"Michael Ventris found something like that, back in the Fifties," Hubert Penrose's voice broke in from directly behind her.

She turned her head. The colonel was standing by the archaeologists' table; Captain Field and the airdyne pilot had gone out.

"He found a lot of Greek inventories of military stores," Penrose continued. "They were in Cretan Linear B script, and at the head of each list was a little picture, a sword or a helmet or a cooking tripod or a chariot wheel. That's what gave him the key to the script."

"Colonel's getting to be quite an archaeologist," Fitzgerald commented. "We're all learning each others' specialties, on this expedition."

"I heard about that long before this expedition was even contemplated." Penrose was tapping a cigarette on his gold case. "I heard about that back before the Thirty Days' War, at Intelligence School, when I was a lieutenant. As a feat of cryptanalysis, not an archaeological discovery."

"Yes, cryptanalysis," von Ohlmhorst pounced. "The reading of a known language in an unknown form of writing. Ventris' lists were in the known language, Greek. Neither he nor anybody else ever read a word of the Cretan language until the finding of the Greek-Cretan bilingual in 1963, because only with a bilingual text, one language already known, can an unknown ancient language be learned. And what hope, I ask you, have we of finding anything like that here? Martha, you've been working on these Martian texts ever since we landed here—for the last six months. Tell me, have you found a single word to which you can positively assign a meaning?"

"Yes, I think I have one." She was trying hard not to sound too exultant. "*Doma*. It's the name of one of the months of the Martian calendar."

"Where did you find that?" von Ohlmhorst asked. "And how did you establish—?"

"Here." She picked up the photostat and handed it along the table to him. "I'd call this the title page of a magazine."

He was silent for a moment, looking at it. "Yes. I would say so, too. Have you any of the rest of it?"

"I'm working on the first page of the first article, listed there. Wait till I see; yes, here's all I found, together, here." She told him where she had gotten it. "I just gathered it up, at the time, and gave it to Geoffrey and Rosita to photostat; this is the first I've really examined it."

The old man got to his feet, brushing tobacco ashes from the front of his jacket, and came to where she was sitting, laying the title page on the table and leafing quickly through the stack of photostats.

"Yes, and here is the second article, on page eight, and here's the next one." He finished the pile of photostats. "A couple of pages missing at the end of the last article. This is remarkable; surprising that a thing like a magazine would have survived so long."

"Well, this silicone stuff the Martians used for paper is pretty durable," Hubert Penrose said. "There doesn't seem to have been any water or any other fluid in it originally, so it wouldn't dry out with time."

"Oh, it's not remarkable that the material would have survived. We've found a good many books and papers in excellent condition. But only a really vital culture, an organized culture, will publish magazines, and this civilization had been dying for hundreds of years before the end. It might have been a thousand years before the time they died out completely that such activities as publishing ended."

"Well, look where I found it; in a closet in a cellar. Tossed in there and forgotten, and then ignored when they were stripping the building. Things like that happen."

Penrose had picked up the title page and was looking at it.

"I don't think there's any doubt about this being a magazine, at all." He looked again at the title, his lips moving silently. "*Mastharnorvod Tadavas Sornhulva*. Wonder what it means. But you're right about the date—*Doma* seems to be the name of a month. Yes, you have a word, Dr. Dane."

Sid Chamberlain, seeing that something unusual was going on, had come over from the table at which he was working. After examining the title page and some of the inside pages, he began whispering into the stenophone he had taken from his belt.

"Don't try to blow this up to anything big, Sid," she cautioned. "All we have is the name of a month, and Lord only knows how long it'll be till we even find out which month it was."

"Well, it's a start, isn't it?" Penrose argued. "Grotefend only had the word for 'king' when he started reading Persian cuneiform."

"But I don't have the word for month; just the name of a month. Everybody knew the names of the Persian kings, long before Grotefend."

"That's not the story," Chamberlain said. "What the public back on Terra will be interested in is finding out that the Martians published magazines, just like we do. Something familiar; make the Martians seem more real. More human."

Three men had come in, and were removing their masks and helmets and oxy-tanks, and peeling out of their quilted coveralls. Two were Space Force lieutenants; the third was a youngish civilian with close-cropped blond hair, in a checked woolen shirt. Tony Lattimer and his helpers.

"Don't tell me Martha finally got something out of that stuff?" he asked, approaching the table. He might have been commenting on the antics of the village half-wit, from his tone.

"Yes; the name of one of the Martian months." Hubert Penrose went on to explain, showing the photostat.

Tony Lattimer took it, glanced at it, and dropped it on the table.

"Sounds plausible, of course, but just an assumption. That word may not be the name of a month, at all—could mean 'published' or 'authorized' or 'copyrighted' or anything like that. Fact is, I don't think it's more than a wild guess that that thing's anything like a periodical." He dismissed the subject and turned to Penrose. "I picked out the next building to enter; that tall one with the conical thing on top. It ought to be in pretty good shape inside; the conical top wouldn't allow dust to accumulate, and from the outside nothing seems to be caved in or crushed. Ground level's higher than the other one, about the seventh floor. I found a good place and drilled for the shots; tomorrow I'll blast a hole in it, and if you can spare some people to help, we can start exploring it right away."

"Yes, of course, Dr. Lattimer. I can spare about a dozen, and I suppose you can find a few civilian volunteers," Penrose told him. "What will you need in the way of equipment?"

"Oh, about six demolition-packets; they can all be shot together. And the usual thing in the way of lights, and breaking and digging tools, and climbing equipment in case we run into broken or doubtful stairways. We'll divide into two parties. Nothing ought to be entered for the first time without a qualified archaeologist along. Three parties, if Martha can tear herself away from this catalogue of systematized incomprehensibilities she's making long enough to do some real work."

She felt her chest tighten and her face become stiff. She was pressing her lips together to lock in a furious retort when Hubert Penrose answered for her.

"Dr. Dane's been doing as much work, and as important work, as you have," he said brusquely. "More important work, I'd be inclined to say."

Von Ohlmhorst was visibly distressed; he glanced once toward Sid Chamberlain, then looked hastily away from him. Afraid of a story of dissension among archaeologists getting out.

"Working out a system of pronunciation by which the Martian language could be transliterated was a most important contribution," he said. "And Martha did that almost unassisted."

"Unassisted by Dr. Lattimer, anyway," Penrose added. "Captain Field and Lieutenant Koremitsu did some work, and I helped out a little, but nine-tenths of it she did herself."

"Purely arbitrary," Lattimer disdained. "Why, we don't even know that the Martians could make the same kind of vocal sounds we do."

"Oh, yes, we do," Ivan Fitzgerald contradicted, safe on his own ground. "I haven't seen any actual Martian skulls—these people seem to have been very tidy about disposing of their dead—but from statues and busts and pictures I've seen. I'd say that their vocal organs were identical with our own."

"Well, grant that. And grant that it's going to be impressive to rattle off the names of Martian notables whose statues we find, and that if we're ever able to attribute any placenames, they'll sound a lot better than this horse-doctors' Latin the old astronomers splashed all over the map of Mars," Lattimer said. "What I object to is her wasting time on this stuff, of which nobody will ever be able to read a word if she fiddles around with those lists till there's another hundred feet of loess on this city, when there's so much real work to be done and we're as shorthanded as we are."

That was the first time that had come out in just so many words. She was glad Lattimer had said it and not Selim von Ohlmhorst.

"What you mean," she retorted, "is that it doesn't have the publicity value that digging up statues has."

For an instant, she could see that the shot had scored. Then Lattimer, with a side glance at Chamberlain, answered:

"What I mean is that you're trying to find something that any archaeologist, yourself included, should know doesn't exist. I don't object to your gambling your professional reputation and making a laughing stock of yourself; what I object to is that the blunders of one archaeologist discredit the whole subject in the eyes of the public."

That seemed to be what worried Lattimer most. She was framing a reply when the communication-outlet whistled shrilly, and then squawked: "Cocktail time! One hour to dinner; cocktails in the library, Hut Four!"

The library, which was also lounge, recreation room, and general gathering-place, was already crowded; most of the crowd was at the long table topped with sheets of glasslike plastic that had been wall panels out of one of the ruined buildings. She poured herself what passed, here, for a martini, and carried it over to where Selim von Ohlmhorst was sitting alone.

For a while, they talked about the building they had just finished exploring, then drifted into reminiscences of their work on Terra—von Ohlmhorst's in Asia Minor, with the Hittite Empire, and hers in Pakistan, excavating the cities of the Harappa Civilization. They finished their drinks—the ingredients were plentiful; alcohol and flavoring extracts synthesized from Martian vegetation—and von Ohlmhorst took the two glasses to the table for refills.

"You know, Martha," he said, when he returned, "Tony was right about one thing. You are gambling your professional standing and reputation. It's against all archaeological experience that a language so completely dead as this one could be deciphered. There was a continuity between all the other ancient languages—by knowing Greek, Champollion learned to read Egyptian; by knowing Egyptian, Hittite was learned. That's why you and your colleagues have never been able to translate the Harappa hieroglyphics; no such continuity exists there. If you insist that this utterly dead language can be read, your reputation will suffer for it."

"I heard Colonel Penrose say, once, that an officer who's afraid to risk his military reputation seldom makes much of a reputation. It's the same with us. If we really want to find things out, we have to risk making mistakes. And I'm a lot more interested in finding things out than I am in my reputation."

She glanced across the room, to where Tony Lattimer was sitting with Gloria Standish, talking earnestly, while Gloria sipped one of the counterfeit martinis and listened. Gloria was the leading contender for the title of Miss Mars, 1996, if you liked big bosomy blondes, but Tony would have been just as attentive to her if she'd looked like the Wicked Witch in "The Wizard of Oz." because Gloria was the Pan-Federation Telecast System commentator with the expedition.

"I know you are," the old Turco-German was saying. "That's why, when they asked me to name another archaeologist for this expedition, I named you."

He hadn't named Tony Lattimer; Lattimer had been pushed onto the expedition by his university. There'd been a lot of high-level string-pulling to that; she wished she knew the whole story. She'd managed to keep clear of universities and university politics; all her digs had been sponsored by non-academic foundations or art museums.

"You have an excellent standing: much better than my own, at your age. That's why it disturbs me to see you jeopardizing it by this insistence that the Martian language can be translated. I can't, really, see how you can hope to succeed."

She shrugged and drank some more of her cocktail, then lit another cigarette. It was getting tiresome to try to verbalize something she only felt.

"Neither do I, now, but I will. Maybe I'll find something like the picture-books Sachiko was talking about. A child's primer, maybe; surely they had things like that. And if I don't. I'll find something else. We've only been here six months. I can wait the rest of my life, if I have to, but I'll do it sometime."

"I can't wait so long," von Ohlmhorst said. "The rest of my life will only be a few years, and when the *Schiaparelli* orbits in, I'll be going back to Terra on the *Cyrano*."

"I wish you wouldn't. This is a whole new world of archaeology. Literally."

"Yes." He finished the cocktail and looked at his pipe as though wondering whether to re-light it so soon before dinner, then put it in his pocket. "A whole new world—but I've grown old, and it isn't for me. I've spent my life studying the Hittites. I can speak the Hittite language, though maybe King Muwatallis wouldn't be able to understand my modern Turkish accent. But the things I'd have to learn here—chemistry, physics, engineering, how to run analytic tests on steel girders and beryllo-silver alloys and plastics and silicones. I'm more at home with a civilization that rode in chariots and fought with swords and was just learning how to work iron. Mars is for young people. This expedition is a cadre of leadership—not only the Space Force people, who'll be the commanders of the main expedition, but us scientists, too. And I'm just an old cavalry general who can't learn to command tanks and aircraft. You'll have time to learn about Mars. I won't."

His reputation as the dean of Hittitologists was solid and secure, too, she added mentally. Then she felt ashamed of the thought. He wasn't to be classed with Tony Lattimer.

"All I came for was to get the work started," he was continuing. "The Federation Government felt that an old hand should do that. Well, it's started, now; you and Tony and whoever come out on the *Schiaparelli* must carry it on. You said it, yourself; you have a whole new world. This is only one city, of the last Martian civilization. Behind this, you have the Late Upland Culture, and the Canal Builders, and all the civilizations and races and empires before them, clear back to the Martian Stone Age." He hesitated for a moment. "You have no idea what all you have to learn, Martha. This isn't the time to start specializing too narrowly."

They all got out of the truck and stretched their legs and looked up the road to the tall building with the queer conical cap askew on its top. The four little figures that had been busy against its wall climbed into the jeep and started back slowly, the smallest of them, Sachiko Koremitsu, paying out an electric cable behind. When it pulled up beside the truck, they climbed out; Sachiko attached the free end of the cable to a nuclear-electric battery. At once, dirty gray smoke and orange dust puffed out from the wall of the building, and, a second later, the multiple explosion banged.

She and Tony Lattimer and Major Lindemann climbed onto the truck, leaving the jeep stand by the road. When they reached the building, a satisfyingly wide breach had been blown in the wall. Lattimer had placed his shots between two of the windows; they were both blown out along with the wall between, and lay unbroken on the ground. Martha remembered the first building they had entered. A Space Force officer had picked up a stone and thrown it at one of the windows, thinking that would be all they'd need to do. It had bounced back. He had drawn his pistol—they'd all carried guns, then, on the principle that what they didn't know about Mars might easily hurt them—and fired four shots. The bullets had ricocheted , screaming thinly; there were four coppery smears of jacket-metal on the window, and a little surface spalling. Somebody tried a rifle; the 4000-f.s. bullet had cracked the glasslike pane without penetrating. An oxyacetylene torch had taken an hour to cut the window out; the lab crew, aboard the ship, were still trying to find out just what the stuff was.

Tony Lattimer had gone forward and was sweeping his flashlight back and forth, swearing petulantly, his voice harshened and amplified by his helmet-speaker.

"I thought I was blasting into a hallway; this lets us into a room. Careful; there's about a two-foot drop to the floor, and a lot of rubble from the blast just inside."

He stepped down through the breach; the others began dragging equipment out of the trucks—shovels and picks and crowbars and sledges, portable floodlights, cameras, sketching materials, an extension ladder, even Alpinists' ropes and crampons and pickaxes. Hubert Penrose was shouldering something that looked like a surrealist machine gun but which was really a nuclear-electric jack-hammer. Martha selected one of the spike-shod mountaineer's ice axes, with which she could dig or chop or poke or pry or help herself over rough footing.

The windows, grimed and crusted with fifty millennia of dust, filtered in a dim twilight; even the breach in the wall, in the morning shade, lighted only a small patch of floor. Somebody snapped on a floodlight, aiming it at the ceiling. The big room was empty and bare; dust lay thick on the floor and reddened the once-white walls. It could have been a large office, but there was nothing left in it to indicate its use.

"This one's been stripped up to the seventh floor!" Lattimer exclaimed. "Street level'll be cleaned out, completely."

"Do for living quarters and shops, then," Lindemann said. "Added to the others, this'll take care of everybody on the *Schiaparelli*."

"Seem to have been a lot of electric or electronic apparatus over along this wall," one of the Space Force officers commented. "Ten or twelve electric outlets." He brushed the dusty wall with his glove, then scraped on the floor with his foot. "I can see where things were pried loose."

The door, one of the double sliding things the Martians had used, was closed. Selim von Ohlmhorst tried it, but it was stuck fast. The metal latch-parts had frozen together, molecule bonding itself to molecule, since the door had last been closed. Hubert Penrose came over with the jack-hammer, fitting a spear-point chisel into place. He set the chisel in the joint between the doors, braced the hammer against his hip, and squeezed the trigger-switch. The hammer banged briefly like the weapon it resembled, and the doors popped a few inches apart, then stuck. Enough dust had worked into the recesses into which it was supposed to slide to block it on both sides.

That was old stuff; they ran into that every time they had to force a door, and they were prepared for it. Somebody went outside and brought in a power-jack and finally one of the doors inched back to the door jamb. That was enough to get the lights and equipment through: they all passed from the room to the hallway beyond. About half the other doors were open; each had a number and a single word, *Darfhulva,* over it.

One of the civilian volunteers, a woman professor of natural ecology from Penn State University, was looking up and down the hall.

"You know," she said, "I feel at home here. I think this was a college of some sort, and these were classrooms. That word, up there; that was the subject taught, or the department. And those electronic devices, all where the class would face them; audio-visual teaching aids."

"A twenty-five-story university?" Lattimer scoffed. "Why, a building like this would handle thirty thousand students."

"Maybe there were that many. This was a big city, in its prime," Martha said, moved chiefly by a desire to oppose Lattimer.

"Yes, but think of the snafu in the halls, every time they changed classes. It'd take half an hour to get everybody back and forth from one floor to another." He turned to von Ohlmhorst. "I'm going up above this floor. This place has been looted clean up to here, but there's a chance there may be something above," he said.

"I'll stay on this floor, at present," the Turco-German replied. "There will be much coming and going, and dragging things in and out. We should get this completely examined and recorded first. Then Major Lindemann's people can do their worst, here."

"Well, if nobody else wants it, I'll take the downstairs," Martha said.

"I'll go along with you," Hubert Penrose told her. "If the lower floors have no archaeological value, we'll turn them into living quarters. I like this building: it'll give everybody room to keep out from under everybody else's feet." He looked down the hall. "We ought to find escalators at the middle."

The hallway, too, was thick underfoot with dust. Most of the open rooms were empty, but a few contained furniture, including small seat-desks. The original proponent of the university theory pointed these out as just what might be found in classrooms. There were escalators, up and down, on either side of the hall, and more on the intersecting passage to the right.

"That's how they handled the students, between classes," Martha commented. "And I'll bet there are more ahead, there."

They came to a stop where the hallway ended at a great square central hall. There were elevators, there, on two of the sides, and four escalators, still usable as stairways. But it was the walls, and the paintings on them, that brought them up short and staring.

They were clouded with dirt—she was trying to imagine what they must have looked like originally, and at the same time estimating the labor that would be involved in cleaning them—but they were still distinguishable, as was the word, *Darfhulva*, in golden letters above each of the four sides. It was a moment before she realized, from the murals, that she had at last found a meaningful Martian word. They were a vast historical panorama, clockwise around the room. A group of skin-clad savages squatting around a fire. Hunters with bows and spears, carrying a carcass of an animal slightly like a pig. Nomads riding long-legged, graceful mounts like hornless deer. Peasants sowing and reaping; mud-walled hut villages, and cities; processions of priests and warriors; battles with swords and bows, and with cannon and muskets; galleys, and ships with sails, and ships without visible means of propulsion, and aircraft. Changing costumes and weapons and machines and styles of architecture. A richly fertile landscape, gradually merging into barren deserts and bushlands—the time of the great planet-wide drought. The Canal Builders—men with machines recognizable as steam-shovels and derricks, digging and quarrying and driving across the empty plains with aqueducts . More cities—seaports on the shrinking oceans; dwindling, half-deserted cities; an abandoned city, with four tiny humanoid figures and a thing like a combat-car in the middle of a brush-grown plaza, they and their vehicle dwarfed by the huge lifeless buildings around them. She had not the least doubt; *Darfhulva* was History.

"Wonderful!" von Ohlmhorst was saying. "The entire history of this race. Why, if the painter depicted appropriate costumes and weapons and machines for each period, and got the architecture right, we can break the history of this planet into eras and periods and civilizations."

"You can assume they're authentic. The faculty of this university would insist on authenticity in the *Darfhulva*—History—Department," she said.

"Yes! *Darfhulva*—History! And your magazine was a journal of *Sornhulva*!" Penrose exclaimed. "You have a word, Martha!" It took her an instant to realize that he had called her by her first name, and not Dr. Dane. She wasn't sure if that weren't a bigger triumph than learning a word of the Martian language. Or a more auspicious start. "Alone, I suppose that *hulva* means something like science or knowledge, or study; combined, it would be equivalent to our 'ology. And *darf* would mean something like past, or old times, or human events, or chronicles."

"That gives you three words, Martha!" Sachiko jubilated. "You did it."

"Let's don't go too fast," Lattimer said, for once not derisively. "I'll admit that *darfhulva* is the Martian word for history as a subject of study; I'll admit that *hulva* is the general word and *darf* modifies it and tells us which subject is meant. But as for assigning specific meanings, we can't do that because we don't know just how the Martians thought, scientifically or otherwise."

He stopped short, startled by the blue-white light that blazed as Sid Chamberlain's Kliegettes went on. When the whirring of the camera stopped, it was Chamberlain who was speaking:

"This is the biggest thing yet; the whole history of Mars, stone age to the end, all on four walls. I'm taking this with the fast shutter, but we'll telecast it in slow motion, from the beginning to the end. Tony, I want you to do the voice for it—running commentary, interpretation of each scene as it's shown. Would you do that?"

Would he do that! Martha thought. If he had a tail, he'd be wagging it at the very thought.

"Well, there ought to be more murals on the other floors," she said. "Who wants to come downstairs with us?"

Sachiko did; immediately. Ivan Fitzgerald volunteered. Sid decided to go upstairs with Tony Lattimer, and Gloria Standish decided to go upstairs, too. Most of the party would remain on the seventh floor, to help Selim von Ohlmhorst get it finished. After poking tentatively at the escalator with the spike of her ice axe, Martha led the way downward.

The sixth floor was *Darfhulva*, too; military and technological history, from the character of the murals. They looked around the central hall, and went down to the fifth; it was like the floors above except that the big quadrangle was stacked with dusty furniture and boxes. Ivan Fitzgerald, who was carrying the floodlight, swung it slowly around. Here the murals were of heroic-sized Martians, so human in appearance as to seem members of her own race, each holding some object—a book, or a test tube, or some bit of scientific apparatus, and behind them were scenes of laboratories and factories, flame and smoke, lightning-flashes. The word at the top of each of the four walls was one with which she was already familiar—*Sornhulva.*

"Hey, Martha; there's that word," Ivan Fitzgerald exclaimed. "The one in the title of your magazine." He looked at the paintings. "Chemistry, or physics."

"Both." Hubert Penrose considered. "I don't think the Martians made any sharp distinction between them. See, the old fellow with the scraggly whiskers must be the inventor of the spectroscope; he has one in his hands, and he has a rainbow behind him. And the woman in the blue smock, beside him, worked in organic chemistry; see the diagrams of long-chain molecules behind her. What word would convey the idea of chemistry and physics taken as one subject?"

"*Sornhulva,*" Sachiko suggested. "If *hulva's* something like science, "*sorn*" must mean matter, or substance, or physical object. You were right, all along, Martha. A civilization like this would certainly leave something like this, that would be self-explanatory."

"This'll wipe a little more of that superior grin off Tony Lattimer's face," Fitzgerald was saying, as they went down the motionless escalator to the floor below. "Tony wants to be a big shot. When you want to be a big shot, you can't bear the possibility of anybody else being a bigger big shot, and whoever makes a start on reading this language will be the biggest big shot archaeology ever saw."

That was true. She hadn't thought of it, in that way, before, and now she tried not to think about it. She didn't want to be a big shot. She wanted to be able to read the Martian language, and find things out about the Martians.

Two escalators down, they came out on a mezzanine around a wide central hall on the street level, the floor forty feet below them and the ceiling thirty feet above. Their lights picked out object after object below—a huge group of sculptured figures in the middle; some kind of a motor vehicle jacked up on trestles for repairs; things that looked like machine-guns and auto-cannon; long tables, tops littered with a dust-covered miscellany; machinery; boxes and crates and containers.

They made their way down and walked among the clutter, missing a hundred things for every one they saw, until they found an escalator to the basement. There were three basements, one under another, until at last they stood at the bottom of the last escalator, on a bare concrete floor, swinging the portable floodlight over stacks of boxes and barrels and drums, and heaps of powdery dust. The boxes were plastic—nobody had ever found anything made of wood in the city—and the barrels and drums were of metal or glass or some glasslike substance. They were outwardly intact. The powdery heaps might have been anything organic, or anything containing fluid. Down here, where wind and dust could not reach, evaporation had been the only force of destruction after the minute life that caused putrefaction had vanished.

They found refrigeration rooms, too, and using Martha's ice axe and the pistollike vibratool Sachiko carried on her belt, they pounded and pried one open, to find dessicated piles of what had been vegetables, and leathery chunks of meat. Samples of that stuff, rocketed up to the ship, would give a reliable estimate, by radio-carbon dating, of how long ago this building had been occupied. The refrigeration unit, radically different from anything their own culture had produced, had been electrically powered. Sachiko and Penrose, poking into it, found the switches still on; the machine had only ceased to function when the power-source, whatever that had been, had failed.

The middle basement had also been used, at least toward the end, for storage; it was cut in half by a partition pierced by but one door. They took half an hour to force this, and were on the point of sending above for heavy equipment when it yielded enough for them to squeeze through. Fitzgerald, in the lead with the light, stopped short, looked around, and then gave a groan that came through his helmet-speaker like a foghorn.

"Oh, no! *No!*"

"What's the matter, Ivan?" Sachiko, entering behind him, asked anxiously.

He stepped aside. "Look at it, Sachi! Are we going to have to do all that?"

Martha crowded through behind her friend and looked around, then stood motionless, dizzy with excitement. Books. Case on case of books, half an acre of cases, fifteen feet to the ceiling. Fitzgerald, and Penrose, who had pushed in behind her, were talking in rapid excitement; she only heard the sound of their voices, not their words. This must be the main stacks of the university library—the entire literature of the vanished race of Mars. In the center, down an aisle between the cases, she could see the hollow square of the librarians' desk, and stairs and a dumb-waiter to the floor above.

She realized that she was walking forward, with the others, toward this. Sachiko was saying: "I'm the lightest; let me go first." She must be talking about the spidery metal stairs.

"I'd say they were safe," Penrose answered. "The trouble we've had with doors around here shows that the metal hasn't deteriorated."

In the end, the Japanese girl led the way, more catlike than ever in her caution. The stairs were quite sound, in spite of their fragile appearance, and they all followed her. The floor above was a duplicate of the room they had entered, and seemed to contain about as many books. Rather than waste time forcing the door here, they returned to the middle basement and came up by the escalator down which they had originally descended.

The upper basement contained kitchens—electric stoves, some with pots and pans still on them—and a big room that must have been, originally, the students' dining room, though when last used it had been a workshop. As they expected, the library reading room was on the street-level floor, directly above the stacks. It seemed to have been converted into a sort of common living room for the building's last occupants. An adjoining auditorium had been made into a chemical works; there were vats and distillation apparatus, and a metal fractionating tower that extended through a hole knocked in the ceiling seventy feet above. A good deal of plastic furniture of the sort they had been finding everywhere in the city was stacked about, some of it broken up, apparently for reprocessing. The other rooms on the street floor seemed also to have been devoted to manufacturing and repair work; a considerable industry, along a number of lines, must have been carried on here for a long time after the university had ceased to function as such.

On the second floor, they found a museum; many of the exhibits remained, tantalizingly half-visible in grimed glass cases. There had been administrative offices there, too. The doors of most of them were closed, and they did not waste time trying to force them, but those that were open had been turned into living quarters. They made notes, and rough floor plans, to guide them in future more thorough examination; it was almost noon before they had worked their way back to the seventh floor.

Selim von Ohlmhorst was in a room on the north side of the building, sketching the position of things before examining them and collecting them for removal. He had the floor checkerboarded with a grid of chalked lines, each numbered.

"We have everything on this floor photographed," he said. "I have three gangs—all the floodlights I have—sketching and making measurements. At the rate we're going, with time out for lunch, we'll be finished by the middle of the afternoon."

"You've been working fast. Evidently you aren't being high-church about a 'qualified archaeologist' entering rooms first," Penrose commented.

"Ach, childishness!" the old man exclaimed impatiently. "These officers of yours aren't fools. All of them have been to Intelligence School and Criminal Investigation School. Some of the most careful amateur archaeologists I ever knew were retired soldiers or policemen. But there isn't much work to be done. Most of the rooms are either empty or like this one—a few bits of furniture and broken trash and scraps of paper. Did you find anything down on the lower floors?"

"Well, yes," Penrose said, a hint of mirth in his voice. "What would you say, Martha?"

She started to tell Selim. The others, unable to restrain their excitement, broke in with interruptions. Von Ohlmhorst was staring in incredulous amazement.

"But this floor was looted almost clean, and the buildings we've entered before were all looted from the street level up," he said, at length.

"The people who looted this one lived here," Penrose replied. "They had electric power to the last; we found refrigerators full of food, and stoves with the dinner still on them. They must have used the elevators to haul things down from the upper floor. The whole first floor was converted into workshops and laboratories. I think that this place must have been something like a monastery in the Dark Ages in Europe, or what such a monastery would have been like if the Dark Ages had followed the fall of a highly developed scientific civilization. For one thing, we found a lot of machine guns and light auto-cannon on the street level, and all the doors were barricaded. The people here were trying to keep a civilization running after the rest of the planet had gone back to barbarism; I suppose they'd have to fight off raids by the barbarians now and then."

"You're not going to insist on making this building into expedition quarters, I hope, colonel?" von Ohlmhorst asked anxiously.

"Oh, no! This place is an archaeological treasure-house. More than that; from what I saw, our technicians can learn a lot, here. But you'd better get this floor cleaned up as soon as you can, though. I'll have the subsurface part, from the sixth floor down, airsealed. Then we'll put in oxygen generators and power units, and get a couple of elevators into service. For the floors above, we can use temporary airsealing floor by floor, and portable equipment; when we have things atmosphered and lighted and heated, you and Martha and Tony Lattimer can go to work systematically and in comfort, and I'll give you all the help I can spare from the other work. This is one of the biggest things we've found yet."

Tony Lattimer and his companions came down to the seventh floor a little later.

"I don't get this, at all," he began, as soon as he joined them. "This building wasn't stripped the way the others were. Always, the procedure seems to have been to strip from the bottom up, but they seem to have stripped the top floors first, here. All but the very top. I found out what that conical thing is, by the way. It's a wind-rotor, and under it there's an electric generator. This building generated its own power."

"What sort of condition are the generators in?" Penrose asked.

"Well, everything's full of dust that blew in under the rotor, of course, but it looks to be in pretty good shape. Hey, I'll bet that's it! They had power, so they used the elevators to haul stuff down. That's just what they did. Some of the floors above here don't seem to have been touched, though." He paused momentarily; back of his oxy-mask, he seemed to be grinning. "I don't know that I ought to mention this in front of Martha, but two floors above—we hit a room—it must have been the reference library for one of the departments—that had close to five hundred books in it."

The noise that interrupted him, like the squawking of a Brobdingnagian parrot, was only Ivan Fitzgerald laughing through his helmet-speaker.

Lunch at the huts was a hasty meal, with a gabble of full-mouthed and excited talking. Hubert Penrose and his chief subordinates snatched their food in a huddled consultation at one end of the table; in the afternoon, work was suspended on everything else and the fifty-odd men and women of the expedition concentrated their efforts on the University. By the middle of the afternoon, the seventh floor had been completely examined, photographed and sketched, and the murals in the square central hall covered with protective tarpaulins, and Laurent Gicquel and his airsealing crew had moved in and were at work. It had been decided to seal the central hall at the entrances. It took the French-Canadian engineer most of the afternoon to find all the ventilation-ducts and plug them. An elevator-shaft on the north side was found reaching clear to the twenty-fifth floor; this would give access to the top of the building; another shaft, from the center, would take care of the floors below. Nobody seemed willing to trust the ancient elevators, themselves; it was the next evening before a couple of cars and the necessary machinery could be fabricated in the machine shops aboard the ship and sent down by landing-rocket. By that time, the airsealing was finished, the nuclear-electric energy-converters were in place, and the oxygen generators set up.

Martha was in the lower basement, an hour or so before lunch the day after, when a couple of Space Force officers came out of the elevator, bringing extra lights with them. She was still using oxygen-equipment; it was a moment before she realized that the newcomers had no masks, and that one of them was smoking. She took off her own helmet-speaker, throat-mike and mask and unslung her tank-pack, breathing cautiously. The air was chilly, and musty-acrid with the odor of antiquity—the first Martian odor she had smelled—but when she lit a cigarette, the lighter flamed clear and steady and the tobacco caught and burned evenly.

The archaeologists, many of the other civilian scientists, a few of the Space Force officers and the two news-correspondents, Sid Chamberlain and Gloria Standish, moved in that evening, setting up cots in vacant rooms. They installed electric stoves and a refrigerator in the old Library Reading Room, and put in a bar and lunch counter. For a few days, the place was full of noise and activity, then, gradually, the Space Force people and all but a few of the civilians returned to their own work. There was still the business of airsealing the more habitable of the buildings already explored, and fitting them up in readiness for the arrival, in a year and a half, of the five hundred members of the main expedition. There was work to be done enlarging the landing field for the ship's rocket craft, and building new chemical-fuel tanks.

There was the work of getting the city's ancient reservoirs cleared of silt before the next spring thaw brought more water down the underground aqueducts everybody called canals in mistranslation of Schiaparelli's Italian word, though this was proving considerably easier than anticipated. The ancient Canal-Builders must have anticipated a time when their descendants would no longer be capable of maintenance work, and had prepared against it. By the day after the University had been made completely habitable, the actual work there was being done by Selim, Tony Lattimer and herself, with half a dozen Space Force officers, mostly girls, and four or five civilians, helping.

They worked up from the bottom, dividing the floor-surfaces into numbered squares, measuring and listing and sketching and photographing. They packaged samples of organic matter and sent them up to the ship for Carbon-14 dating and analysis; they opened cans and jars and bottles, and found that everything fluid in them had evaporated, through the porosity of glass and metal and plastic if there were no other way. Wherever they looked, they found evidence of activity suddenly suspended and never resumed. A vise with a bar of metal in it, half cut through and the hacksaw beside it. Pots and pans with hardened remains of food in them; a leathery cut of meat on a table, with the knife ready at hand. Toilet articles on washstands; unmade beds, the bedding ready to crumble at a touch but still retaining the impress of the sleeper's body; papers and writing materials on desks, as though the writer had gotten up, meaning to return and finish in a fifty-thousand-year-ago moment.

It worried her. Irrationally, she began to feel that the Martians had never left this place; that they were still around her, watching disapprovingly every time she picked up something they had laid down. They haunted her dreams, now, instead of their enigmatic writing. At first, everybody who had moved into the University had taken a separate room, happy to escape the crowding and lack of privacy of the huts. After a few nights, she was glad when Gloria Standish moved in with her, and accepted the newswoman's excuse that she felt lonely without somebody to talk to before falling asleep. Sachiko Koremitsu joined them the next evening, and before going to bed, the girl officer cleaned and oiled her pistol, remarking that she was afraid some rust may have gotten into it.

The others felt it, too. Selim von Ohlmhorst developed the habit of turning quickly and looking behind him, as though trying to surprise somebody or something that was stalking him. Tony Lattimer, having a drink at the bar that had been improvised from the librarian's desk in the Reading Room, set down his glass and swore.

"You know what this place is? It's an archaeological *Marie Celeste*!" he declared. "It was occupied right up to the end—we've all seen the shifts these people used to keep a civilization going here—but what was the end? What happened to them? Where did they go?"

"You didn't expect them to be waiting out front, with a red carpet and a big banner, *Welcome Terrans*, did you, Tony?" Gloria Standish asked.

"No, of course not; they've all been dead for fifty thousand years. But if they were the last of the Martians, why haven't we found their bones, at least? Who buried them, after they were dead?" He looked at the glass, a bubble-thin goblet, found, with hundreds of others like it, in a closet above, as though debating with himself whether to have another drink. Then he voted in the affirmative and reached for the cocktail pitcher. "And every door on the old ground level is either barred or barricaded from the inside. How did they get out? And why did they leave?"

The next day, at lunch, Sachiko Koremitsu had the answer to the second question. Four or five electrical engineers had come down by rocket from the ship, and she had been spending the morning with them, in oxy-masks, at the top of the building.

"Tony, I thought you said those generators were in good shape," she began, catching sight of Lattimer. "They aren't. They're in the most unholy mess I ever saw. What happened, up there, was that the supports of the wind-rotor gave way, and weight snapped the main shaft, and smashed everything under it."

"Well, after fifty thousand years, you can expect something like that," Lattimer retorted. "When an archaeologist says something's in good shape, he doesn't necessarily mean it'll start as soon as you shove a switch in."

"You didn't notice that it happened when the power was on, did you," one of the engineers asked, nettled at Lattimer's tone. "Well, it was. Everything's burned out or shorted or fused together; I saw one busbar eight inches across melted clean in two. It's a pity we didn't find things in good shape, even archaeologically speaking. I saw a lot of interesting things, things in advance of what we're using now. But it'll take a couple of years to get everything sorted out and figure what it looked like originally."

"Did it look as though anybody'd made any attempt to fix it?" Martha asked.

Sachiko shook her head. "They must have taken one look at it and given up. I don't believe there would have been any possible way to repair anything."

"Well, that explains why they left. They needed electricity for lighting, and heating, and all their industrial equipment was electrical. They had a good life, here, with power; without it, this place wouldn't have been habitable."

"Then why did they barricade everything from the inside, and how did they get out?" Lattimer wanted to know.

"To keep other people from breaking in and looting. Last man out probably barred the last door and slid down a rope from upstairs," von Ohlmhorst suggested. "This Houdini-trick doesn't worry me too much. We'll find out eventually."

"Yes, about the time Martha starts reading Martian," Lattimer scoffed.

"That may be just when we'll find out," von Ohlmhorst replied seriously. "It wouldn't surprise me if they left something in writing when they evacuated this place."

"Are you really beginning to treat this pipe dream of hers as a serious possibility, Selim?" Lattimer demanded. "I know, it would be a wonderful thing, but wonderful things don't happen just because they're wonderful. Only because they're possible, and this isn't. Let me quote that distinguished Hittitologist, Johannes Friedrich: 'Nothing can be translated out of nothing.' Or that later but not less distinguished Hittitologist, Selim von Ohlmhorst: 'Where are you going to get your bilingual?'"

"Friedrich lived to see the Hittite language deciphered and read," von Ohlmhorst reminded him.

"Yes, when they found Hittite-Assyrian bilinguals." Lattimer measured a spoonful of coffee-powder into his cup and added hot water. "Martha, you ought to know, better than anybody, how little chance you have. You've been working for years in the Indus Valley; how many words of Harappa have you or anybody else ever been able to read?"

"We never found a university, with a half-million-volume library, at Harappa or Mohenjo-Daro."

"And, the first day we entered this building, we established meanings for several words," Selim von Ohlmhorst added.

"And you've never found another meaningful word since," Lattimer added. "And you're only sure of general meaning, not specific meaning of word-elements, and you have a dozen different interpretations for each word."

"We made a start," von Ohlmhorst maintained. "We have Grotefend's word for 'king.' But I'm going to be able to read some of those books, over there, if it takes me the rest of my life here. It probably will, anyhow."

"You mean you've changed your mind about going home on the *Cyrano*?" Martha asked. "You'll stay on here?"

The old man nodded. "I can't leave this. There's too much to discover. The old dog will have to learn a lot of new tricks, but this is where my work will be, from now on."

Lattimer was shocked. "You're nuts!" he cried. "You mean you're going to throw away everything you've accomplished in Hittitology and start all over again here on Mars? Martha, if you've talked him into this crazy decision, you're a criminal!"

"Nobody talked me into anything," von Ohlmhorst said roughly. "And as for throwing away what I've accomplished in Hittitology, I don't know what the devil you're talking about. Everything I know about the Hittite Empire is published and available to anybody. Hittitology's like Egyptology; it's stopped being research and archaeology and become scholarship and history. And I'm not a scholar or a historian; I'm a pick-and-shovel field archaeologist—a highly skilled and specialized grave-robber and junk-picker—and there's more pick-and-shovel work on this planet than I could do in a hundred lifetimes. This is something new; I was a fool to think I could turn my back on it and go back to scribbling footnotes about Hittite kings."

"You could have anything you wanted, in Hittitology. There are a dozen universities that'd sooner have you than a winning football team. But no! You have to be the top man in Martiology, too. You can't leave that for anybody else—" Lattimer shoved his chair back and got to his feet, leaving the table with an oath that was almost a sob of exasperation.

Maybe his feelings were too much for him. Maybe he realized, as Martha did, what he had betrayed. She sat, avoiding the eyes of the others, looking at the ceiling, as embarrassed as though Lattimer had flung something dirty on the table in front of them. Tony Lattimer had, desperately, wanted Selim to go home on the *Cyrano*. Martiology was a new field; if Selim entered it, he would bring with him the reputation he had already built in Hittitology, automatically stepping into the leading role that Lattimer had coveted for himself. Ivan Fitzgerald's words echoed back to her—when you want to be a big shot, you can't bear the possibility of anybody else being a bigger big shot. His derision of her own efforts became comprehensible, too. It wasn't that he was convinced that she would never learn to read the Martian language. He had been afraid that she would.

Ivan Fitzgerald finally isolated the germ that had caused the Finchley girl's undiagnosed illness. Shortly afterward, the malady turned into a mild fever, from which she recovered. Nobody else seemed to have caught it. Fitzgerald was still trying to find out how the germ had been transmitted.

They found a globe of Mars, made when the city had been a seaport. They located the city, and learned that its name had been Kukan—or something with a similar vowel-consonant ratio. Immediately, Sid Chamberlain and Gloria Standish began giving their telecasts a Kukan dateline, and Hubert Penrose used the name in his official reports. They also found a Martian calendar; the year had been divided into ten more or less equal months, and one of them had been Doma. Another month was Nor, and that was a part of the name of the scientific journal Martha had found.

Bill Chandler, the zoologist, had been going deeper and deeper into the old sea bottom of Syrtis. Four hundred miles from Kukan, and at fifteen thousand feet lower altitude, he shot a bird. At least, it was a something with wings and what were almost but not quite feathers, though it was more reptilian than avian in general characteristics. He and Ivan Fitzgerald skinned and mounted it, and then dissected the carcass almost tissue by tissue. About seven- eighths of its body capacity was lungs; it certainly breathed air containing at least half enough oxygen to support human life, or five times as much as the air around Kukan.

That took the center of interest away from archaeology, and started a new burst of activity. All the expedition's aircraft—four jetticopters and three wingless airdyne reconnaissance fighters—were thrown into intensified exploration of the lower sea bottoms, and the bio-science boys and girls were wild with excitement and making new discoveries on each flight.

The University was left to Selim and Martha and Tony Lattimer, the latter keeping to himself while she and the old Turco-German worked together. The civilian specialists in other fields, and the Space Force people who had been holding tape lines and making sketches and snapping cameras, were all flying to lower Syrtis to find out how much oxygen there was and what kind of life it supported.

Sometimes Sachiko dropped in; most of the time she was busy helping Ivan Fitzgerald dissect specimens. They had four or five species of what might loosely be called birds, and something that could easily be classed as a reptile, and a carnivorous mammal the size of a cat with birdlike claws, and a herbivore almost identical with the piglike thing in the big *Darfhulva* mural, and another like a gazelle with a single horn in the middle of its forehead.

The high point came when one party, at thirty thousand feet below the level of Kukan, found breathable air. One of them had a mild attack of *sorroche* and had to be flown back for treatment in a hurry, but the others showed no ill effects.

The daily newscasts from Terra showed a corresponding shift in interest at home. The discovery of the University had focused attention on the dead past of Mars; now the public was interested in Mars as a possible home for humanity. It was Tony Lattimer who brought archaeology back into the activities of the expedition and the news at home.

Martha and Selim were working in the museum on the second floor, scrubbing the grime from the glass cases, noting contents, and grease-penciling numbers; Lattimer and a couple of Space Force officers were going through what had been the administrative offices on the other side. It was one of these, a young second lieutenant, who came hurrying in from the mezzanine, almost bursting with excitement.

"Hey, Martha! Dr. von Ohlmhorst!" he was shouting. "Where are you? Tony's found the Martians!"

Selim dropped his rag back in the bucket; she laid her clipboard on top of the case beside her.

"Where?" they asked together.

"Over on the north side." The lieutenant took hold of himself and spoke more deliberately. "Little room, back of one of the old faculty offices—conference room. It was locked from the inside, and we had to burn it down with a torch. That's where they are. Eighteen of them, around a long table—"

Gloria Standish, who had dropped in for lunch, was on the mezzanine, fairly screaming into a radiophone extension:

" ... Dozen and a half of them! Well, of course they're dead. What a question! They look like skeletons covered with leather. No, I do not know what they died of. Well, forget it; I don't care if Bill Chandler's found a three-headed hippopotamus. Sid, don't you get it? We've found the *Martians*!"

She slammed the phone back on its hook, rushing away ahead of them.

Martha remembered the closed door; on the first survey, they hadn't attempted opening it. Now it was burned away at both sides and lay, still hot along the edges, on the floor of the big office room in front. A floodlight was on in the room inside, and Lattimer was going around looking at things while a Space Force officer stood by the door. The center of the room was filled by a long table; in armchairs around it sat the eighteen men and women who had occupied the room for the last fifty millennia. There were bottles and glasses on the table in front of them, and, had she seen them in a dimmer light, she would have thought that they were merely dozing over their drinks. One had a knee hooked over his chair-arm and was curled in foetuslike sleep. Another had fallen forward onto the table, arms extended, the emerald set of a ring twinkling dully on one finger. Skeletons covered with leather, Gloria Standish had called them, and so they were—faces like skulls, arms and legs like sticks, the flesh shrunken onto the bones under it.

"Isn't this something!" Lattimer was exulting. "Mass suicide, that's what it was. Notice what's in the corners?"

Braziers, made of perforated two-gallon-odd metal cans, the white walls smudged with smoke above them. Von Ohlmhorst had noticed them at once, and was poking into one of them with his flashlight.

"Yes; charcoal. I noticed a quantity of it around a couple of hand-forges in the shop on the first floor. That's why you had so much trouble breaking in; they'd sealed the room on the inside." He straightened and went around the room, until he found a ventilator, and peered into it. "Stuffed with rags.

They must have been all that were left, here. Their power was gone, and they were old and tired, and all around them their world was dying. So they just came in here and lit the charcoal, and sat drinking together till they all fell asleep. Well, we know what became of them, now, anyhow."

Sid and Gloria made the most of it. The Terran public wanted to hear about Martians, and if live Martians couldn't be found, a room full of dead ones was the next best thing. Maybe an even better thing; it had been only sixty-odd years since the Orson Welles invasion-scare. Tony Lattimer, the discoverer, was beginning to cash in on his attentions to Gloria and his ingratiation with Sid; he was always either making voice-and-image talks for telecast or listening to the news from the home planet. Without question, he had become, overnight, the most widely known archaeologist in history.

"Not that I'm interested in all this, for myself," he disclaimed, after listening to the telecast from Terra two days after his discovery. "But this is going to be a big thing for Martian archaeology. Bring it to the public attention; dramatize it. Selim, can you remember when Lord Carnarvon and Howard Carter found the tomb of Tutankhamen?"

"In 1923? I was two years old, then," von Ohlmhorst chuckled. "I really don't know how much that publicity ever did for Egyptology. Oh, the museums did devote more space to Egyptian exhibits, and after a museum department head gets a few extra showcases, you know how hard it is to make him give them up. And, for a while, it was easier to get financial support for new excavations. But I don't know how much good all this public excitement really does, in the long run."

"Well, I think one of us should go back on the *Cyrano*, when the *Schiaparelli* orbits in," Lattimer said. "I'd hoped it would be you; your voice would carry the most weight. But I think it's important that one of us go back, to present the story of our work, and what we have accomplished and what we hope to accomplish, to the public and to the universities and the learned societies, and to the Federation Government. There will be a great deal of work that will have to be done. We must not allow the other scientific fields and the so-called practical interests to monopolize public and academic support. So, I believe I shall go back at least for a while, and see what I can do—"

Lectures. The organization of a Society of Martian Archaeology, with Anthony Lattimer, Ph.D., the logical candidate for the chair. Degrees, honors; the deference of the learned, and the adulation of the lay public. Positions, with impressive titles and salaries. Sweet are the uses of publicity.

She crushed out her cigarette and got to her feet. "Well, I still have the final lists of what we found in *Halvhulva*—Biology—department to check over. I'm starting on Sornhulva tomorrow, and I want that stuff in shape for expert evaluation."

That was the sort of thing Tony Lattimer wanted to get away from, the detail-work and the drudgery. Let the infantry do the slogging through the mud; the brass-hats got the medals.

She was halfway through the fifth floor, a week later, and was having midday lunch in the reading room on the first floor when Hubert Penrose came over and sat down beside her, asking her what she was doing. She told him.

"I wonder if you could find me a couple of men, for an hour or so," she added. "I'm stopped by a couple of jammed doors at the central hall. Lecture room and library, if the layout of that floor's anything like the ones below it."

"Yes. I'm a pretty fair door-buster, myself." He looked around the room. "There's Jeff Miles; he isn't doing much of anything. And we'll put Sid Chamberlain to work, for a change, too. The four of us ought to get your doors open." He called to Chamberlain, who was carrying his tray over to the dish washer. "Oh, Sid; you doing anything for the next hour or so?"

"I was going up to the fourth floor, to see what Tony's doing."

"Forget it. Tony's bagged his season limit of Martians. I'm going to help Martha bust in a couple of doors; we'll probably find a whole cemetery full of Martians."

Chamberlain shrugged. "Why not. A jammed door can have anything back of it, and I know what Tony's doing—just routine stuff."

Jeff Miles, the Space Force captain, came over, accompanied by one of the lab-crew from the ship who had come down on the rocket the day before.

"This ought to be up your alley, Mort," he was saying to his companion. "Chemistry and physics department. Want to come along?"

The lab man, Mort Tranter, was willing. Seeing the sights was what he'd come down from the ship for. She finished her coffee and cigarette, and they went out into the hall together, gathered equipment and rode the elevator to the fifth floor.

The lecture hall door was the nearest; they attacked it first. With proper equipment and help, it was no problem and in ten minutes they had it open wide enough to squeeze through with the floodlights. The room inside was quite empty, and, like most of the rooms behind closed doors, comparatively free from dust. The students, it appeared, had sat with their backs to the door, facing a low platform, but their seats and the lecturer's table and equipment had been removed. The two side walls bore inscriptions: on the right, a pattern of concentric circles which she recognized as a diagram of atomic structure, and on the left a complicated table of numbers and words, in two columns. Tranter was pointing at the diagram on the right.

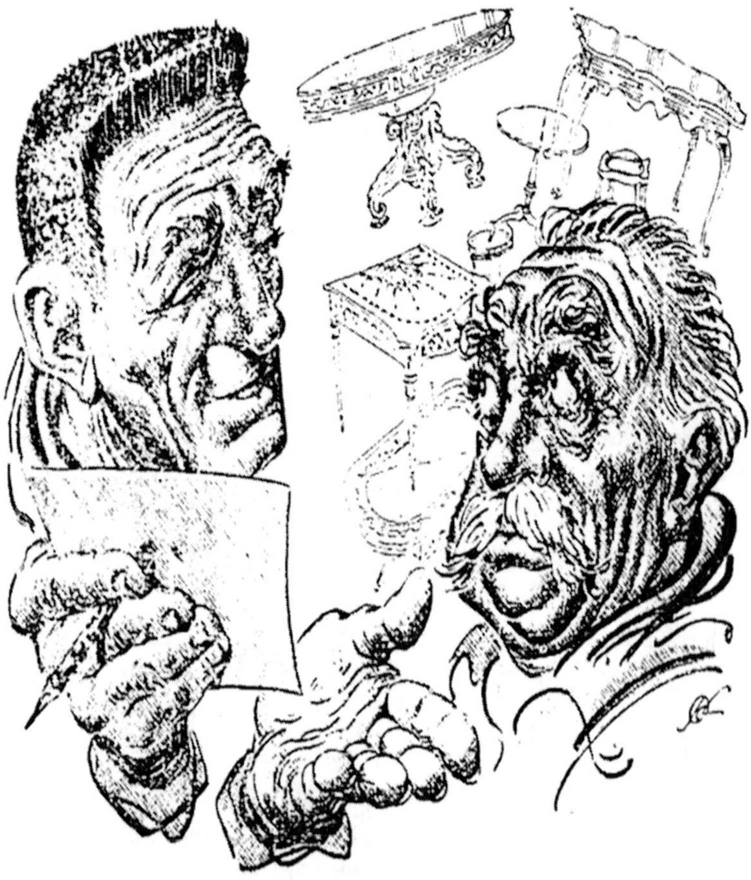

"They got as far as the Bohr atom, anyhow," he said. "Well, not quite. They knew about electron shells, but they have the nucleus pictured as a solid mass. No indication of proton-and-neutron structure. I'll bet, when you come to translate their scientific books, you'll find that they taught that the atom was the ultimate and indivisible particle. That explains why you people never found any evidence that the Martians used nuclear energy."

"That's a uranium atom," Captain Miles mentioned.

"It is?" Sid Chamberlain asked, excitedly. "Then they did know about atomic energy. Just because we haven't found any pictures of A-bomb mushrooms doesn't mean—"

She turned to look at the other wall. Sid's signal reactions were getting away from him again; uranium meant nuclear power to him, and the two words were interchangeable. As she studied the arrangement of the numbers and words, she could hear Tranter saying:

"Nuts, Sid. We knew about uranium a long time before anybody found out what could be done with it. Uranium was discovered on Terra in 1789, by Klaproth."

There was something familiar about the table on the left wall. She tried to remember what she had been taught in school about physics, and what she had picked up by accident afterward. The second column was a continuation of the first: there were forty-six items in each, each item numbered consecutively—

"Probably used uranium because it's the largest of the natural atoms," Penrose was saying. "The fact that there's nothing beyond it there shows that they hadn't created any of the transuranics. A student could go to that thing and point out the outer electron of any of the ninety-two elements."

Ninety-two! That was it; there were ninety-two items in the table on the left wall! Hydrogen was Number One, she knew; One, *Sarfaldsorn*. Helium was Two; that was *Tirfaldsorn*. She couldn't remember which element came next, but in Martian it was *Sarfalddavas*. *Sorn* must mean matter, or substance, then. And *davas*; she was trying to think of what it could be. She turned quickly to the others, catching hold of Hubert Penrose's arm with one hand and waving her clipboard with the other.

"Look at this thing, over here," she was clamoring excitedly. "Tell me what you think it is. Could it be a table of the elements?"

They all turned to look. Mort Tranter stared at it for a moment.

"Could be. If I only knew what those squiggles meant—"

That was right; he'd spent his time aboard the ship.

"If you could read the numbers, would that help?" she asked, beginning to set down the Arabic digits and their Martian equivalents. "It's decimal system, the same as we use."

"Sure. If that's a table of elements, all I'd need would be the numbers. Thanks," he added as she tore off the sheet and gave it to him.

Penrose knew the numbers, and was ahead of him. "Ninety-two items, numbered consecutively. The first number would be the atomic number. Then a single word, the name of the element. Then the atomic weight—"

She began reading off the names of the elements. "I know hydrogen and helium; what's *tirfalddavas*, the third one?"

"Lithium," Tranter said. "The atomic weights aren't run out past the decimal point. Hydrogen's one plus, if that double-hook dingus is a plus sign; Helium's four-plus, that's right. And lithium's given as seven, that isn't right. It's six-point nine-four-oh. Or is that thing a Martian minus sign?"

"Of course! Look! A plus sign is a hook, to hang things together; a minus sign is a knife, to cut something off from something—see, the little loop is the handle and the long pointed loop is the blade. Stylized, of course, but that's what it is. And the fourth element, kiradavas; what's that?"

"Beryllium. Atomic weight given as nine-and-a-hook; actually it's nine-point-oh-two."

Sid Chamberlain had been disgruntled because he couldn't get a story about the Martians having developed atomic energy. It took him a few minutes to understand the newest development, but finally it dawned on him.

"Hey! You're reading that!" he cried. "You're reading Martian!"

"That's right," Penrose told him. "Just reading it right off. I don't get the two items after the atomic weight, though. They look like months of the Martian calendar. What ought they to be, Mort?"

Tranter hesitated. "Well, the next information after the atomic weight ought to be the period and group numbers. But those are words."

"What would the numbers be for the first one, hydrogen?"

"Period One, Group One. One electron shell, one electron in the outer shell," Tranter told her. "Helium's period one, too, but it has the outer—only—electron shell full, so it's in the group of inert elements."

"*Trav, Trav. Trav's* the first month of the year. And helium's *Trav, Yenth; Yenth* is the eighth month."

"The inert elements could be called Group Eight, yes. And the third element, lithium, is Period Two, Group One. That check?"

"It certainly does. *Sanv, Trav; Sanv's* the second month. What's the first element in Period Three?"

"Sodium. Number Eleven."

That's right; it's *Krav, Trav*. Why, the names of the months are simply numbers, one to ten, spelled out.

"*Doma*'s the fifth month. That was your first Martian word, Martha," Penrose told her. "The word for five. And if *davas* is the word for metal, and *sornhulva* is chemistry and / or physics, I'll bet Tadavas Sornhulva is literally translated as: Of-Metal Matter-Knowledge. Metallurgy, in other words. I wonder what Mastharnorvod means." It surprised her that, after so long and with so much happening in the meantime, he could remember that. "Something like 'Journal,' or 'Review,' or maybe 'Quarterly.'"

"We'll work that out, too," she said confidently. After this, nothing seemed impossible. "Maybe we can find—" Then she stopped short. "You said 'Quarterly.' I think it was 'Monthly,' instead. It was dated for a specific month, the fifth one. And if *nor* is ten, Mastharnorvod could be 'Year-Tenth.' And I'll bet we'll find that *masthar* is the word for year." She looked at the table on the wall again. "Well, let's get all these words down, with translations for as many as we can."

"Let's take a break for a minute," Penrose suggested, getting out his cigarettes. "And then, let's do this in comfort. Jeff, suppose you and Sid go across the hall and see what you find in the other room in the way of a desk or something like that, and a few chairs. There'll be a lot of work to do on this."

Sid Chamberlain had been squirming as though he were afflicted with ants, trying to contain himself. Now he let go with an excited jabber.

"This is really it! *The* it, not just it-of-the-week, like finding the reservoirs or those statues or this building, or even the animals and the dead Martians!

Wait till Selim and Tony see this! Wait till Tony sees it; I want to see his face! And when I get this on telecast, all Terra's going to go nuts about it!" He turned to Captain Miles. "Jeff, suppose you take a look at that other door, while I find somebody to send to tell Selim and Tony. And Gloria; wait till she sees this—"

"Take it easy, Sid," Martha cautioned. "You'd better let me have a look at your script, before you go too far overboard on the telecast. This is just a beginning; it'll take years and years before we're able to read any of those books downstairs."

"It'll go faster than you think, Martha," Hubert Penrose told her. "We'll all work on it, and we'll teleprint material to Terra, and people there will work on it. We'll send them everything we can ... everything we work out, and copies of books, and copies of your word-lists—"

And there would be other tables—astronomical tables, tables in physics and mechanics, for instance—in which words and numbers were equivalent. The library stacks, below, would be full of them. Transliterate them into Roman alphabet spellings and Arabic numerals, and somewhere, somebody would spot each numerical significance, as Hubert Penrose and Mort Tranter and she had done with the table of elements. And pick out all the chemistry textbooks in the Library; new words would take on meaning from contexts in which the names of elements appeared. She'd have to start studying chemistry and physics, herself—

Sachiko Koremitsu peeped in through the door, then stepped inside.

"Is there anything I can do—?" she began. "What's happened? Something important?"

"Important?" Sid Chamberlain exploded. "Look at that, Sachi! We're reading it! Martha's found out how to read Martian!" He grabbed Captain Miles by the arm. "Come on, Jeff; let's go. I want to call the others—" He was still babbling as he hurried from the room.

Sachi looked at the inscription. "Is it true?" she asked, and then, before Martha could more than begin to explain, flung her arms around her. "Oh, it really is! You are reading it! I'm so happy!"

She had to start explaining again when Selim von Ohlmhorst entered. This time, she was able to finish.

"But, Martha, can you be really sure? You know, by now, that learning to read this language is as important to me as it is to you, but how can you be so sure that those words really mean things like hydrogen and helium and boron and oxygen? How do you know that their table of elements was anything like ours?"

Tranter and Penrose and Sachiko all looked at him in amazement.

"That isn't just the Martian table of elements; that's *the* table of elements. It's the only one there is." Mort Tranter almost exploded. "Look, hydrogen has one proton and one electron. If it had more of either, it wouldn't be hydrogen, it'd be something else. And the same with all the rest of the elements. And hydrogen on Mars is the same as hydrogen on Terra, or on Alpha Centauri, or in the next galaxy—"

"You just set up those numbers, in that order, and any first-year chemistry student could tell you what elements they represented." Penrose said. "Could if he expected to make a passing grade, that is."

The old man shook his head slowly, smiling. "I'm afraid I wouldn't make a passing grade. I didn't know, or at least didn't realize, that. One of the things I'm going to place an order for, to be brought on the *Schiaparelli*, will be a set of primers in chemistry and physics, of the sort intended for a bright child of ten or twelve. It seems that a Martiologist has to learn a lot of things the Hittites and the Assyrians never heard about."

Tony Lattimer, coming in, caught the last part of the explanation. He looked quickly at the walls and, having found out just what had happened, advanced and caught Martha by the hand.

"You really did it, Martha! You found your bilingual! I never believed that it would be possible; let me congratulate you!"

He probably expected that to erase all the jibes and sneers of the past. If he did, he could have it that way. His friendship would mean as little to her as his derision—except that his friends had to watch their backs and his knife. But he was going home on the *Cyrano*, to be a big shot . Or had this changed his mind for him again?

"This is something we can show the world, to justify any expenditure of time and money on Martian archaeological work. When I get back to Terra, I'll see that you're given full credit for this achievement—"

On Terra, her back and his knife would be out of her watchfulness.

"We won't need to wait that long," Hubert Penrose told him dryly. "I'm sending off an official report, tomorrow; you can be sure Dr. Dane will be given full credit, not only for this but for her previous work, which made it possible to exploit this discovery."

"And you might add, work done in spite of the doubts and discouragements of her colleagues," Selim von Ohlmhorst said. "To which I am ashamed to have to confess my own share."

"You said we had to find a bilingual," she said. "You were right, too."

"This is better than a bilingual, Martha," Hubert Penrose said. "Physical science expresses universal facts; necessarily it is a universal language. Heretofore archaeologists have dealt only with pre-scientific cultures."